AF890424

BIOLOGY WORD WALL

A TEACHER & STUDENT HELPER!

G. Katz Chronicle
Publisher: IMNF Education Press, Pleasant Valley, NY

Library of Congress Control Number: 2022912555

EGESTION

POOPING

FECES

FECAL MATTER

DIFFUSION

PASSIVE TRANSPORT

FACILITATED TRANSPORT

OSMOSIS

INTERIOR
FLUID OF CELLS

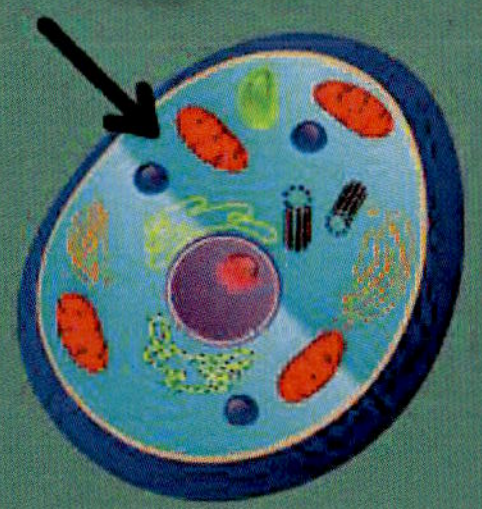

CHROMOSOMES

ARE IN THE
NUCLEUS

AND
CONTAIN
GENES

RELATIONSHIP BETWEEN 2 OR MORE ORGANISMS THAT IS BENEFICIAL TO ALL

ENDOSYMBIOTIC THEORY

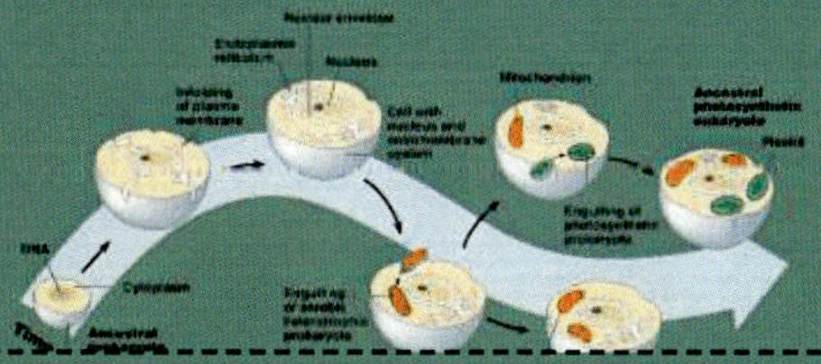

EUKARYOTES EVOLVED FROM PROKARYOTES

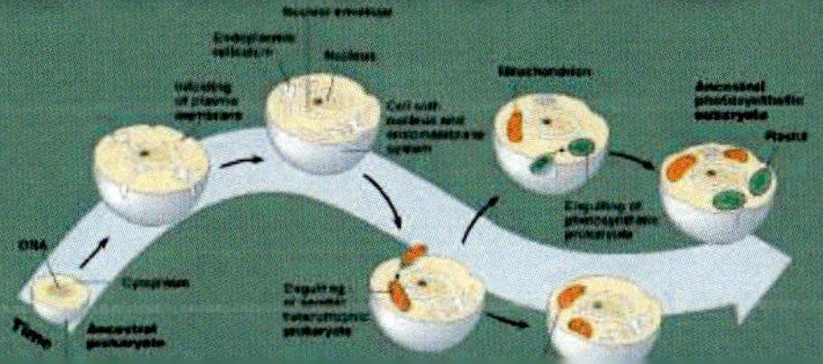

CYTOPLASM

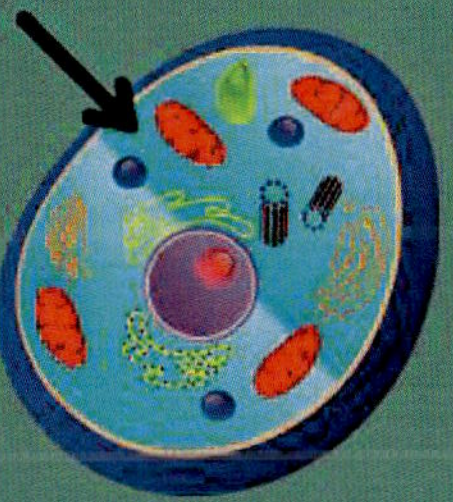

SYMMETRY

BALANCED

THE SAME ON BOTH SIDES

SYMBIOTIC OR SYMBIOSIS

NUCLEAR ENVELOPE

NUCLEAR MEMBRANE

TRANSPORT

CIRCULATION

SELECTIVELY PERMEABLE

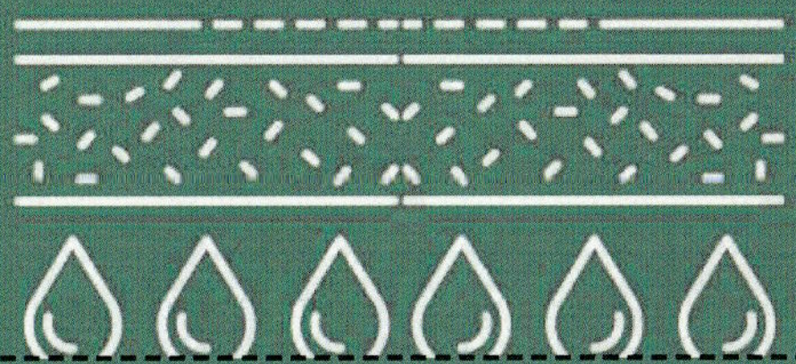

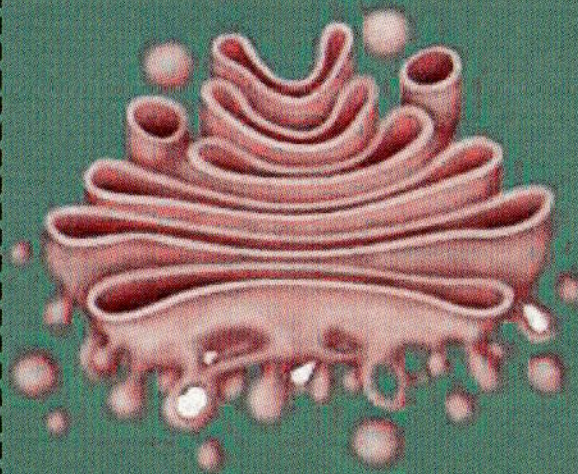

GOLGI APPARATUS

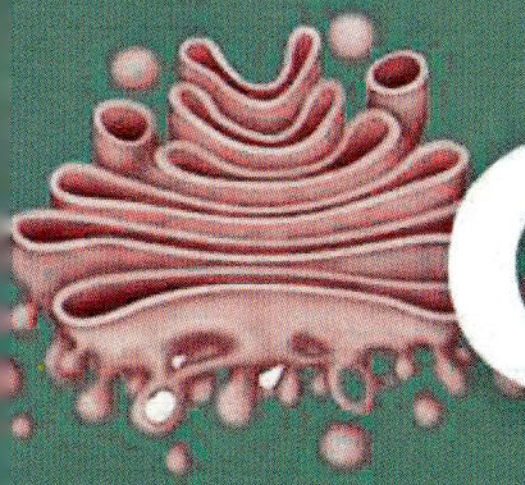

GOLGI BODY

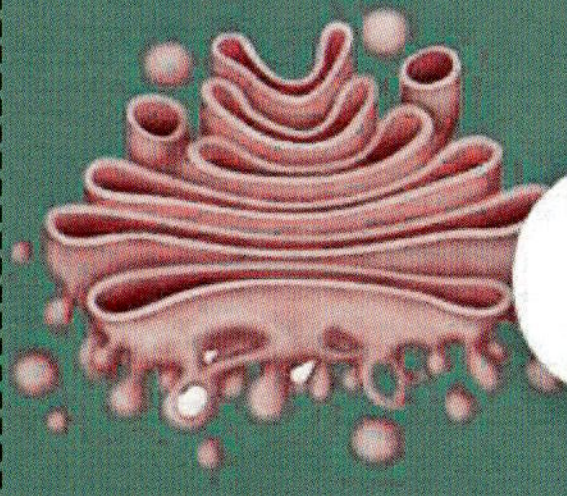

GOLGI COMPLEX

BI-LIPID
LAYER

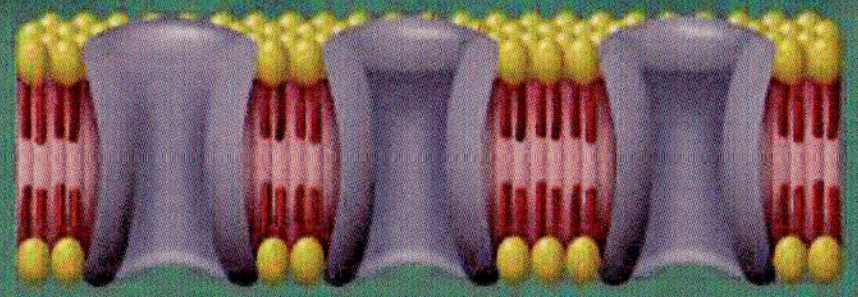

LIPID
BI-LAYER

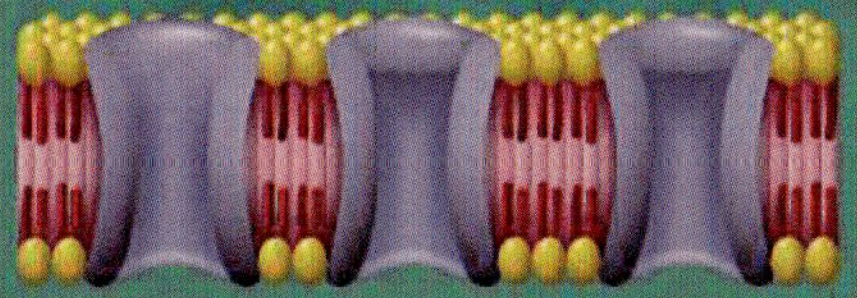

PLASMA
MEMBRANE

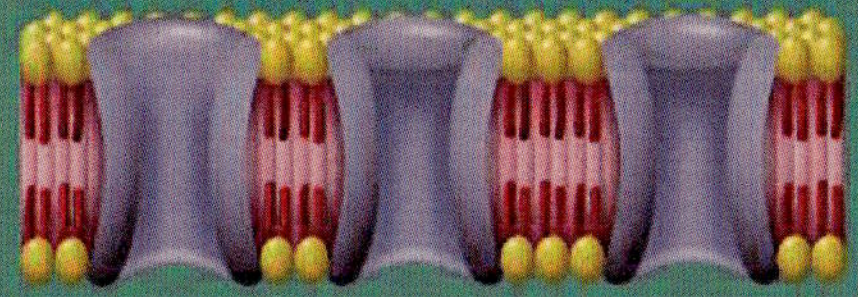

SEMI-
PERMEABLE

TO RESPOND
TO THE
ENVIRONMENT

AND
MAINTAIN
HOMEOSTASIS

CELL
MEMBRANE

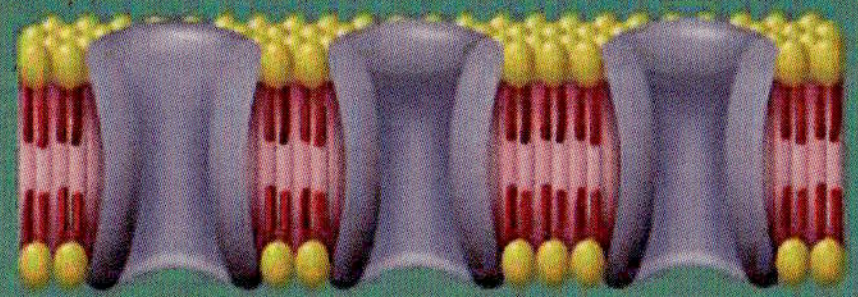

LIPID

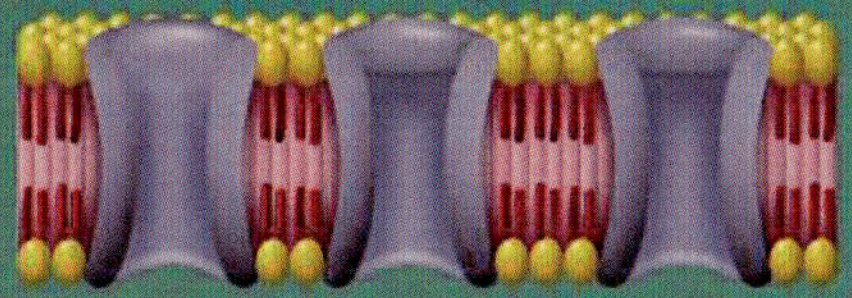

PROTOZOA

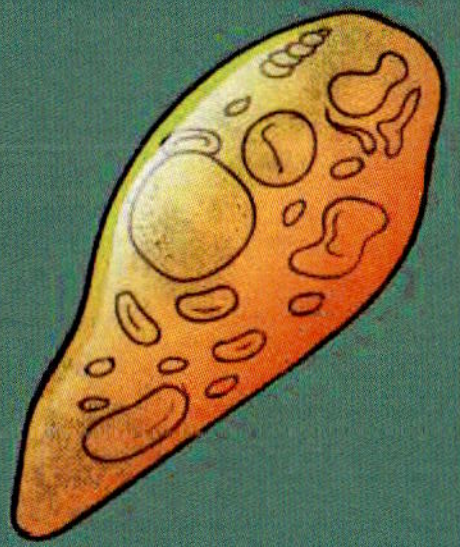

PROTIST

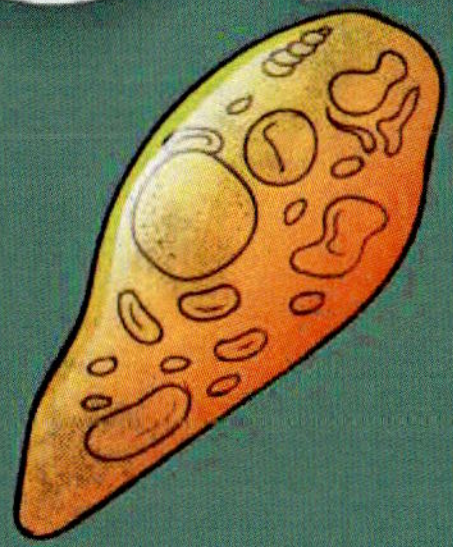

PROTISTA

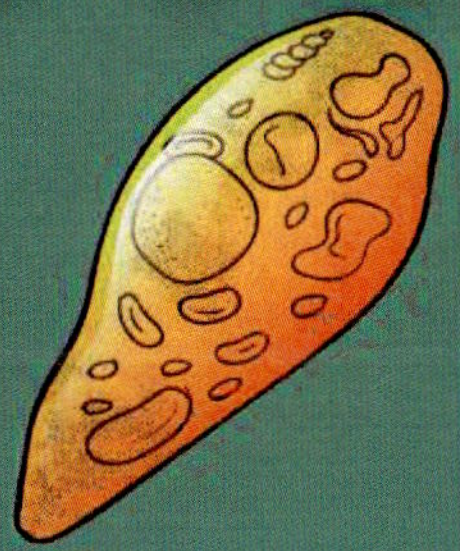

ORGANISMS

PERFORM

REGULATION

LIGHT SOURCE

OCULAR

EYE PIECE

INVESTIGATION

HOMEOSTASIS

INTERNAL BALANCE

ILLUMINATION

DISAGREE

SCIENTIFIC EXPERIMENT

CONTROLLED EXPERIMENT

TEST

OPINION

SUPPORT

AGREE

REFUTE

DEPENDENT VARIABLE

MANIPULATED VARIABLE

GRAPHED ON Y-AXIS / VERTICAL AXIS

BIAS

CONTROL

USED FOR COMPARISON

INDEPENDENT VARIABLE

GRAPHED ON X-AXIS / HORIZONTAL AXIS

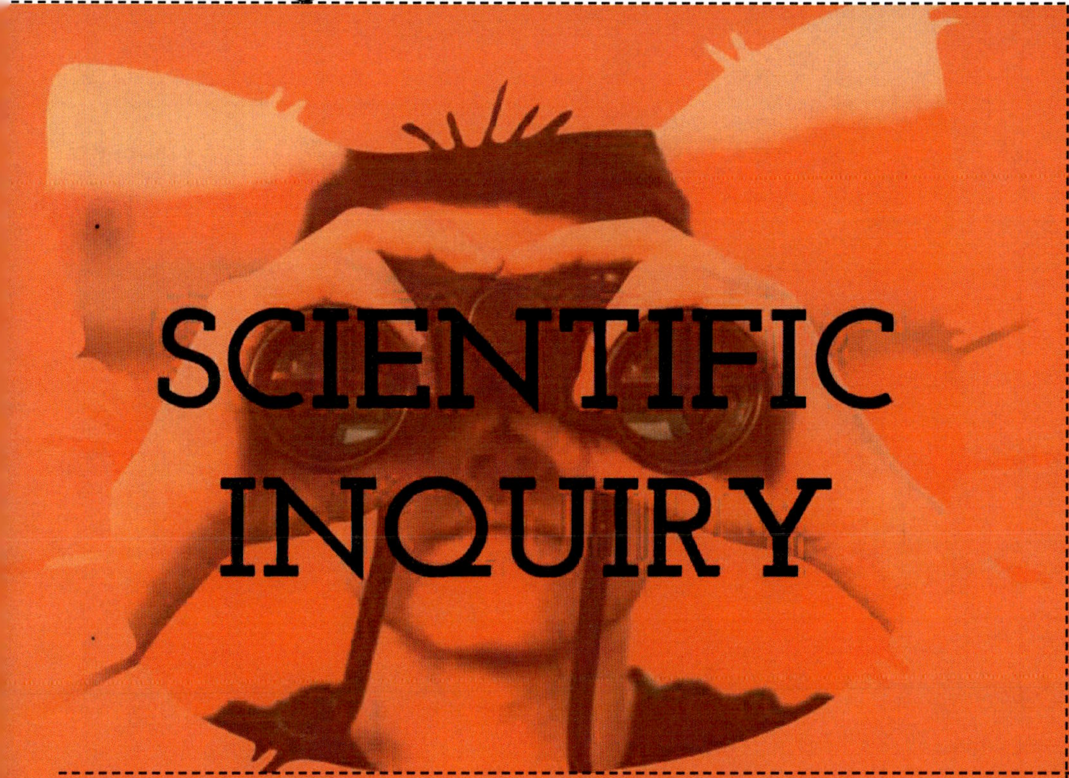

VERIFY

CHECK

REPEATED EXPERIMENTATION

MORULA

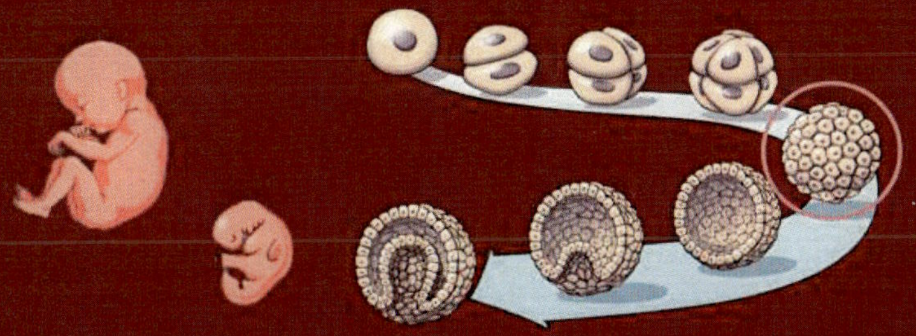

FETUS

EMBRYO

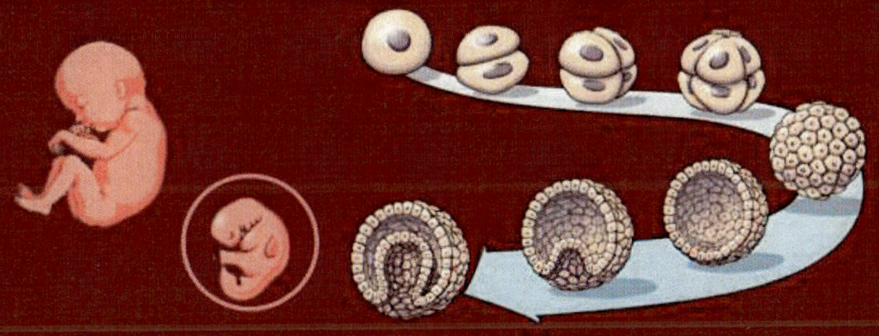

DEVELOPING BABY

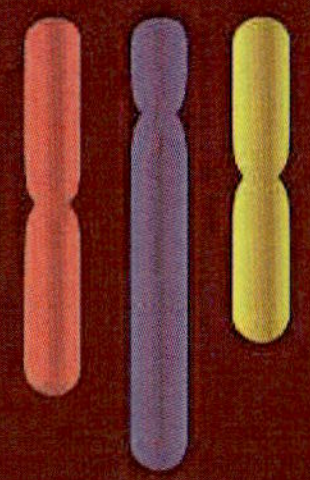

A SET OF UNPAIRED CHROMOSOMES

ZYGOTE

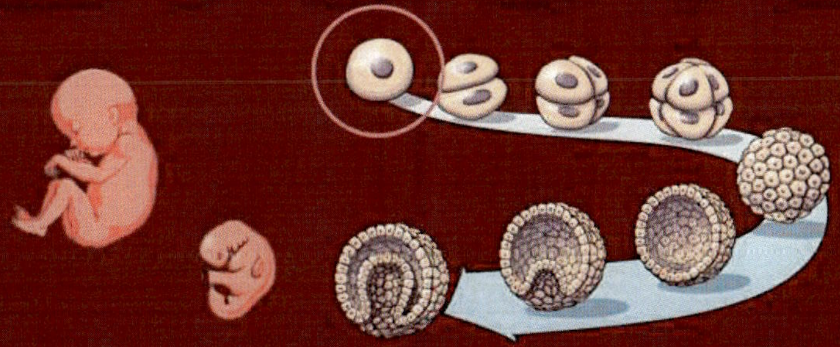

BLASTULA

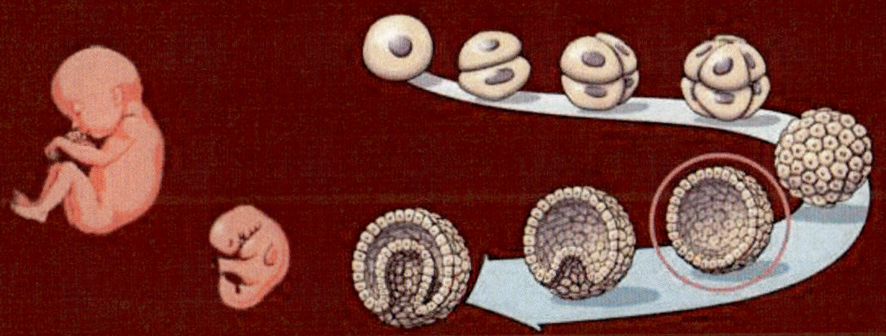

GASTRULA

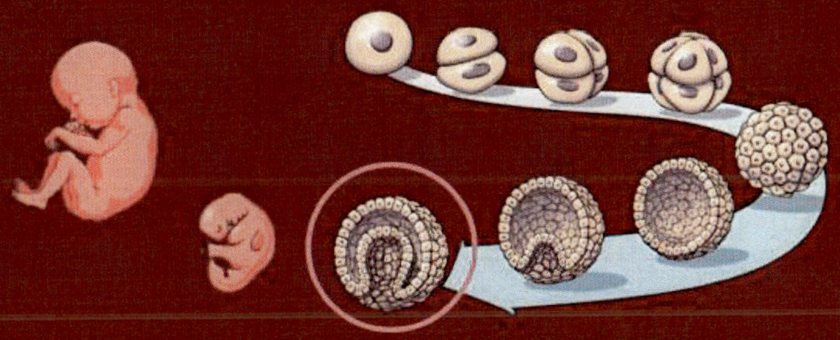

INDEPENDENT ASSORTMENT

CROSSING OVER

HAPLOID

MONOPLOID

CLONE

TWIN

GENE SHUFFLING

DIFFERENT COMBINATIONS OF ALLELES

BUDDING

SPORULATION

REGENERATION

VEGETATIVE PROPAGATION

CHILDREN

BABIES

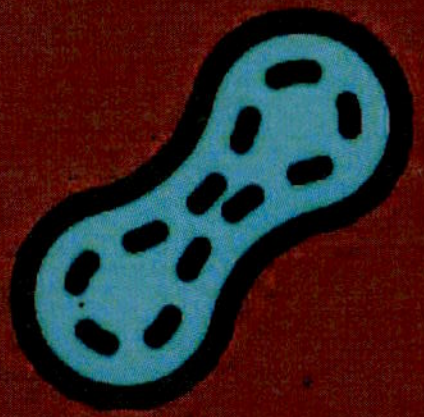

ASEXUAL REPRODUCTION

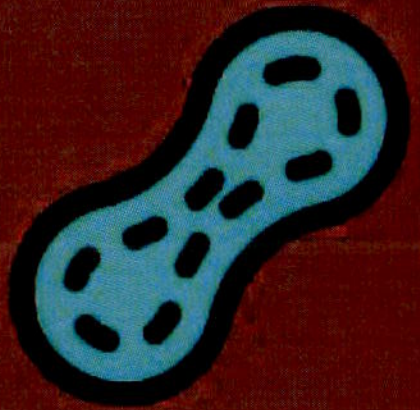

BINARY FISSION

BIRTH CANAL

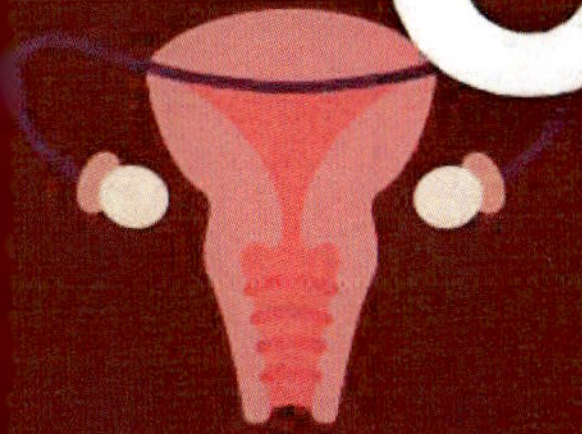

GESTATION

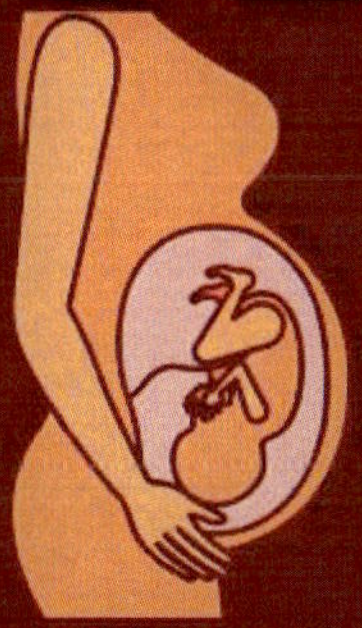

PREGNANCY

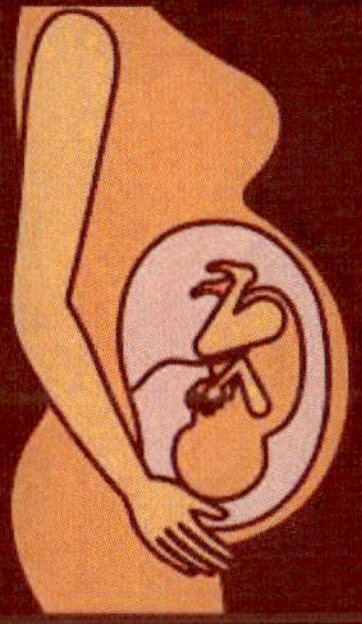

OFFSPRING

SPERM DUCT

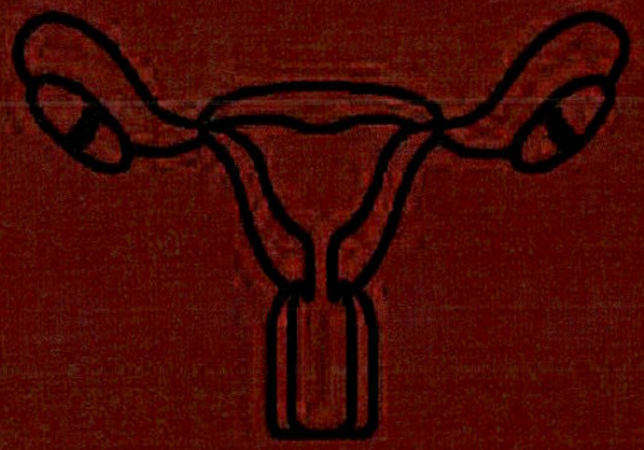

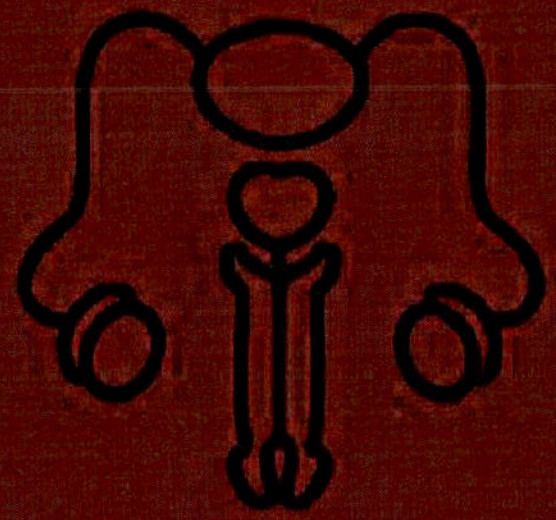

GERMINATE

GROW

FROM

SEED

VAGINA

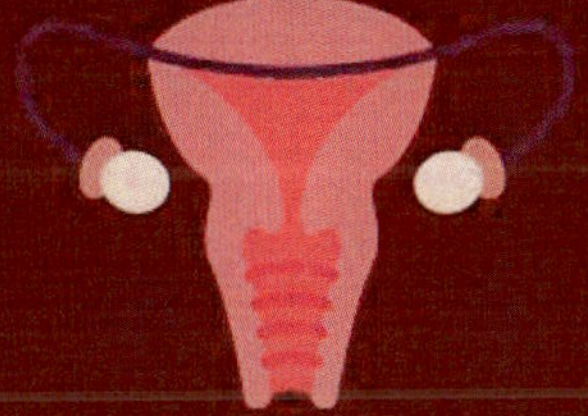

OVIDUCT

FALLOPIAN TUBE

VAS DEFERENCE

DUCTUS DEFERENCE

ASEXUAL REPRODUCTION

MITOSIS

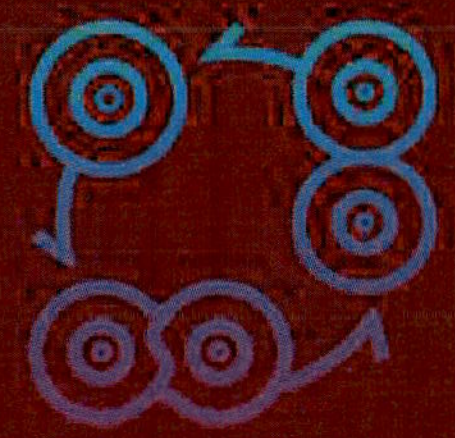

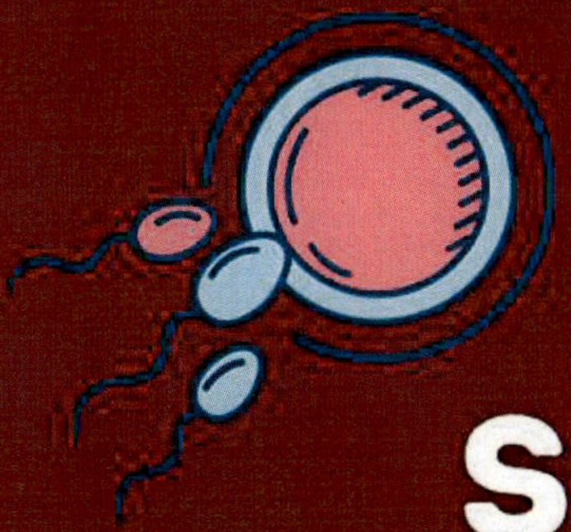

SEXUAL REPRODUCTION

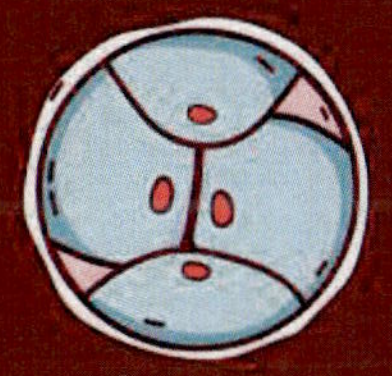

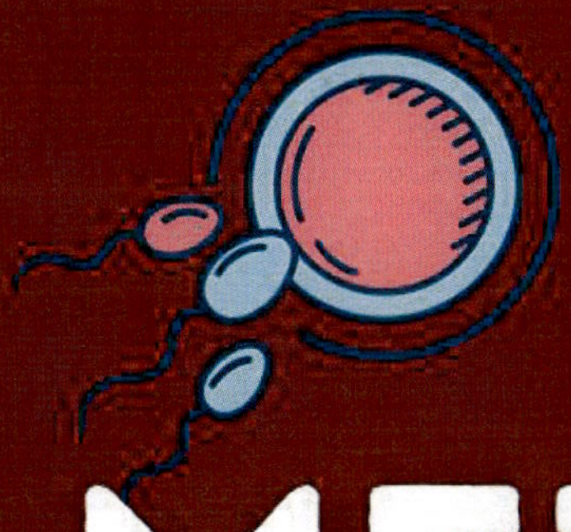

MEIOSIS

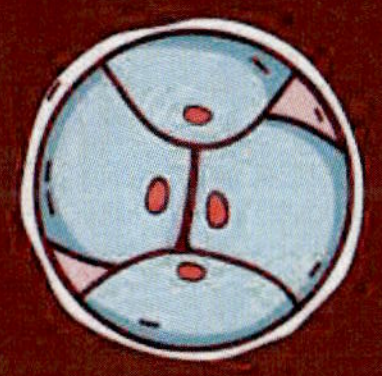

CORRESPONDING

GONADS

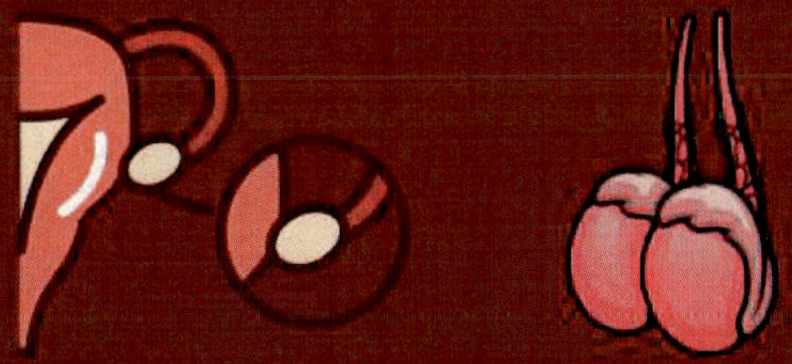

TESTES

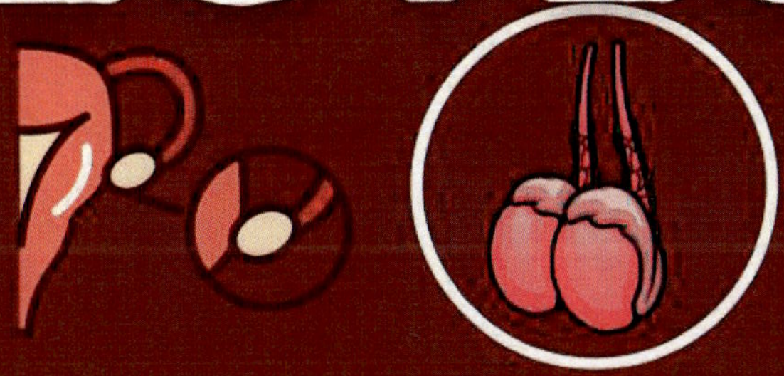

OVARIES

CHROMATIN

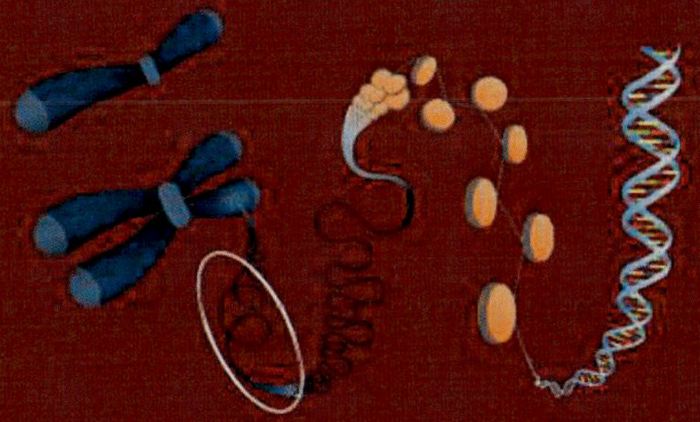

DNA

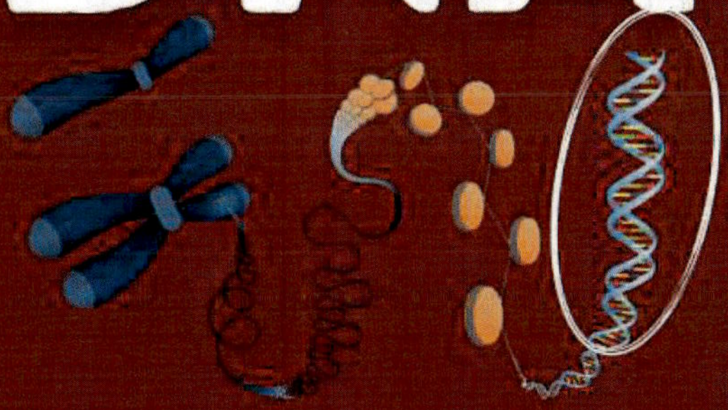

HOMOLOGOUS

SIMILAR

CHROMOSOME

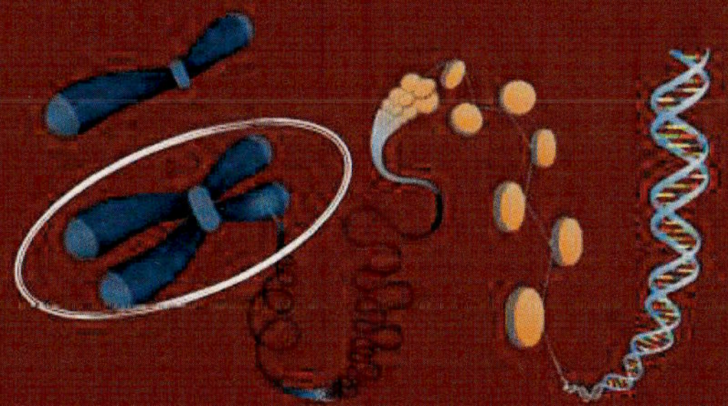

CHROMATID

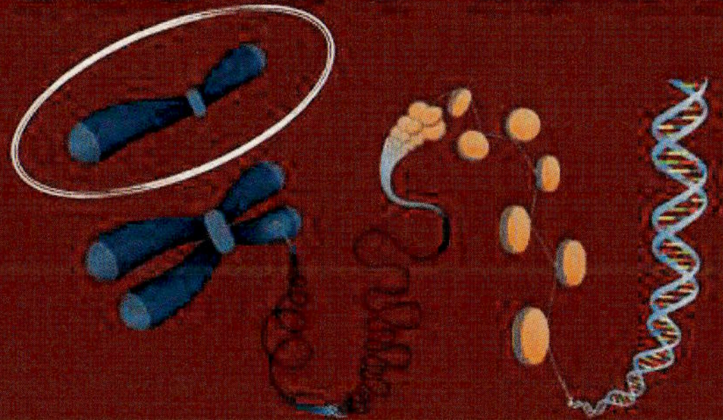

CENTROMERE

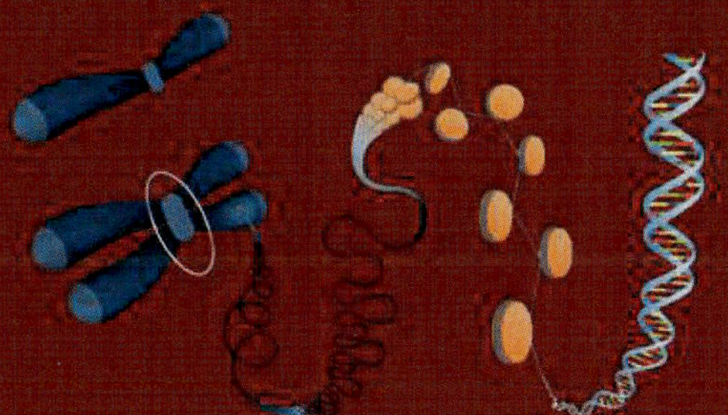

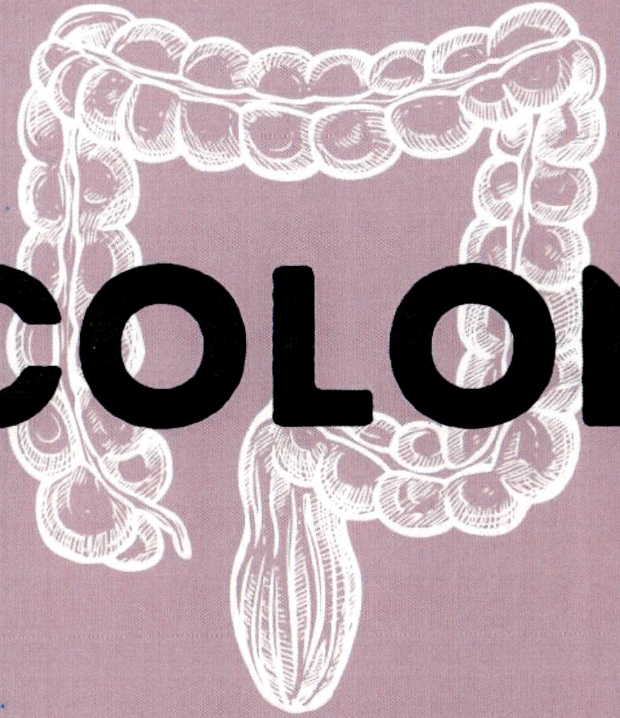

COLON

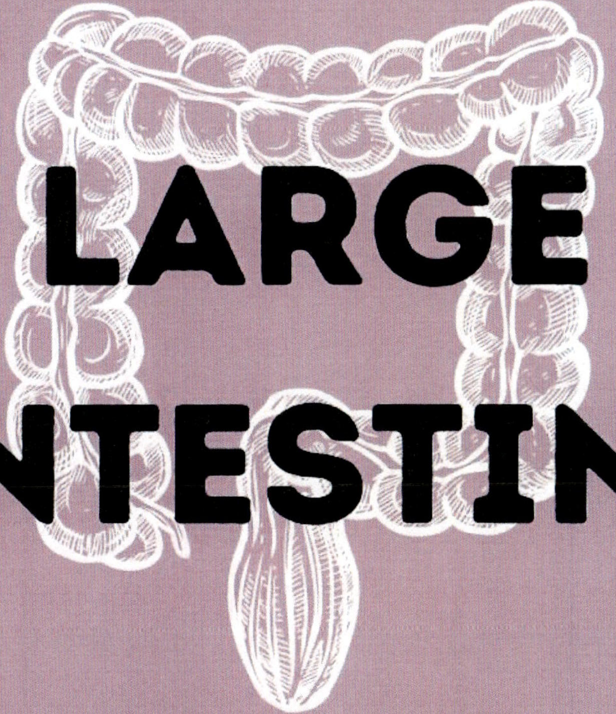

LARGE INTESTINE

NEPHRONS

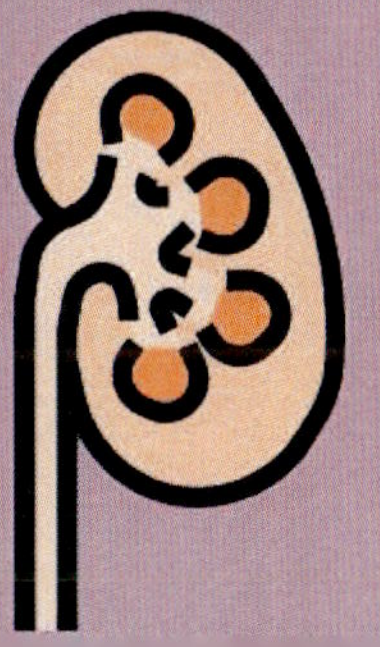

KIDNEY CELLS

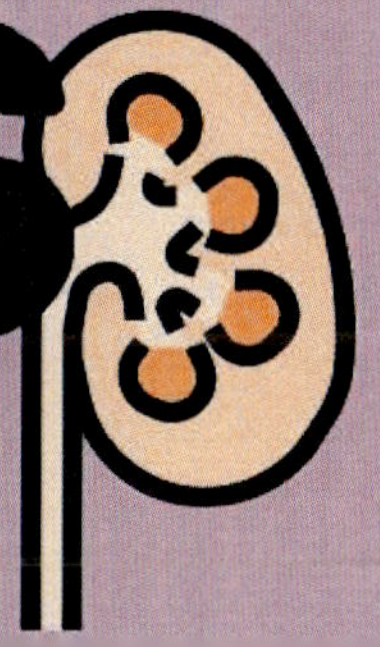

FATIGUE

FEELING TIRED

IMMUNE SYSTEM

LYMPHATIC SYSTEM

CHEMORECEPTOR

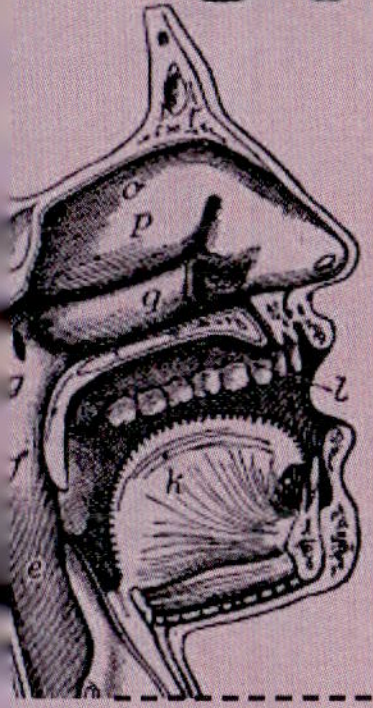

NOSE/MOUTH

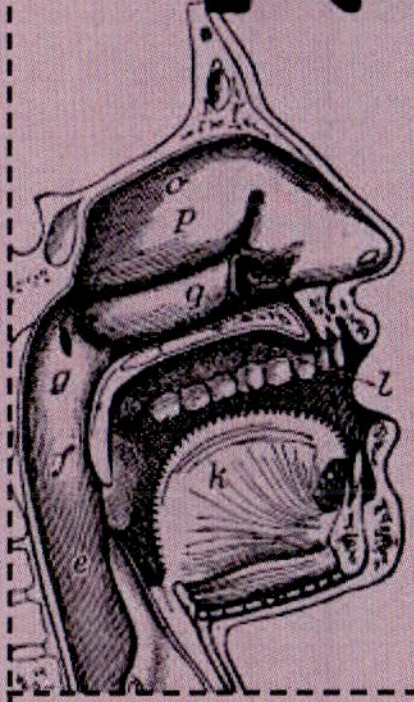

PROPRIOCEPTOR

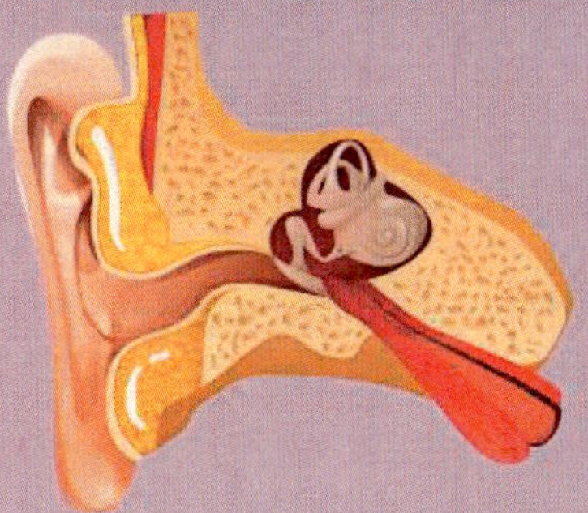

INNER EAR

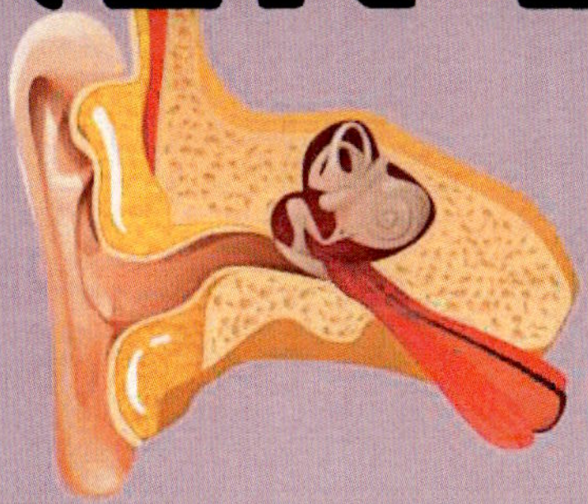

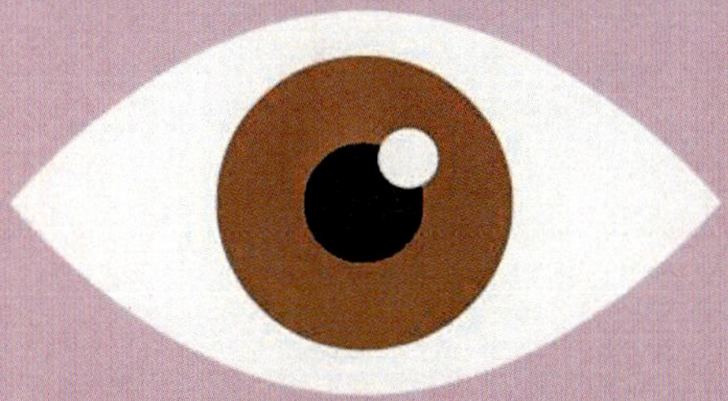

PHOTORECEPTOR

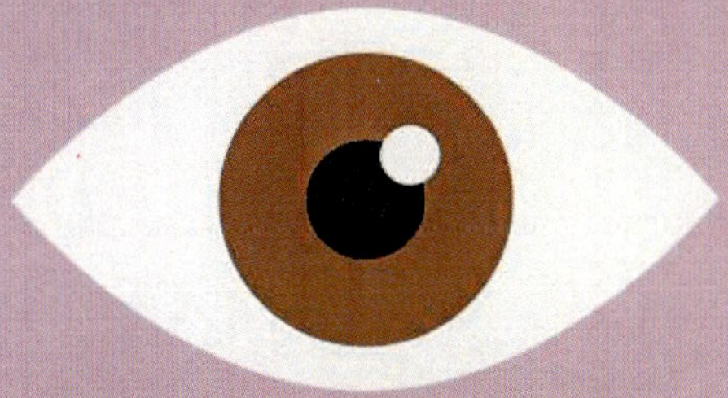

EYE

THERMORECEPTOR

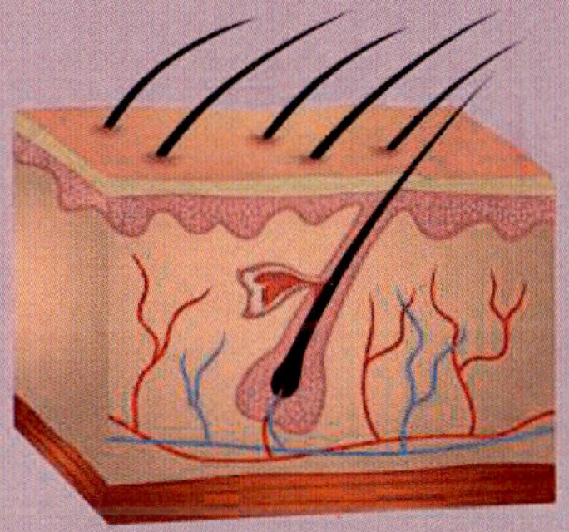

SKIN

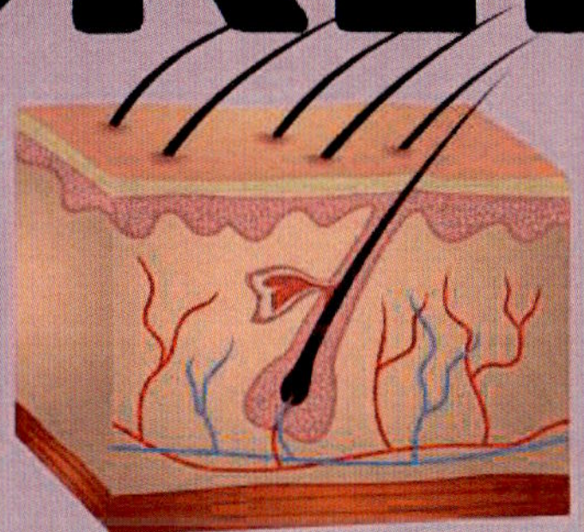

NEURONS

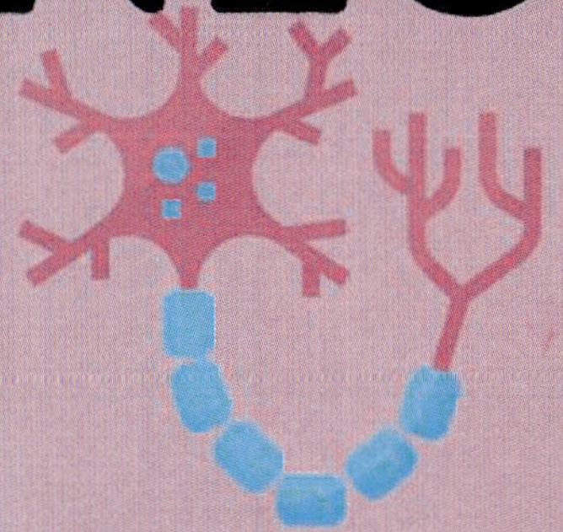

BRAIN CELLS

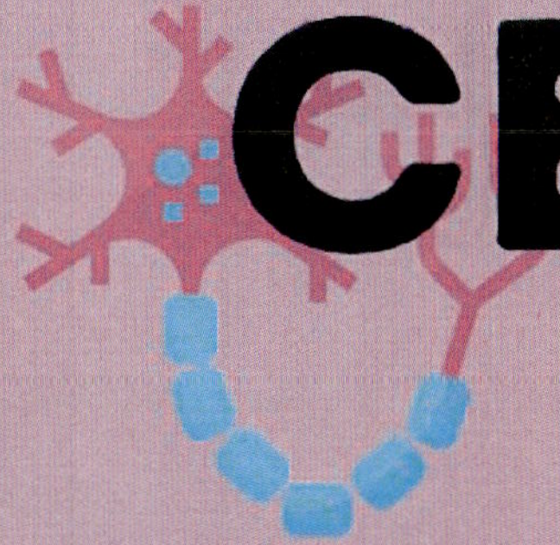

ALVEOLI

LUNG CELLS

SURFACE MARKERS

RECEPTOR MOLECULES

PATHOGEN

CAUSES DISEASE

ANTIBODY

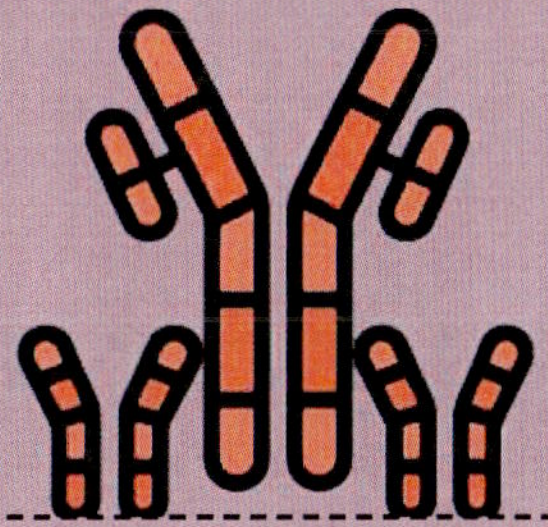

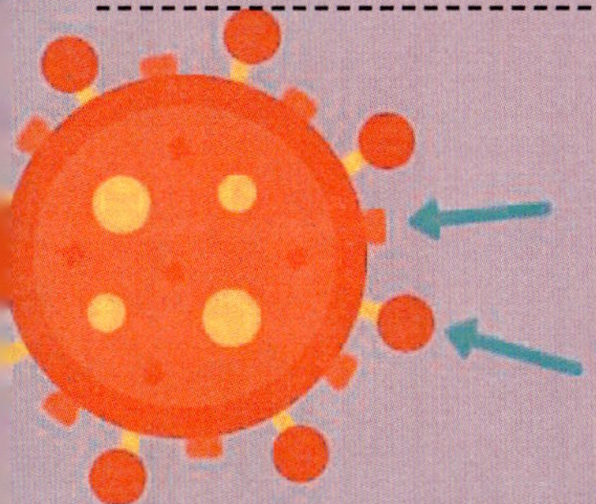

ANTIGEN

INFECTION FIGHTER

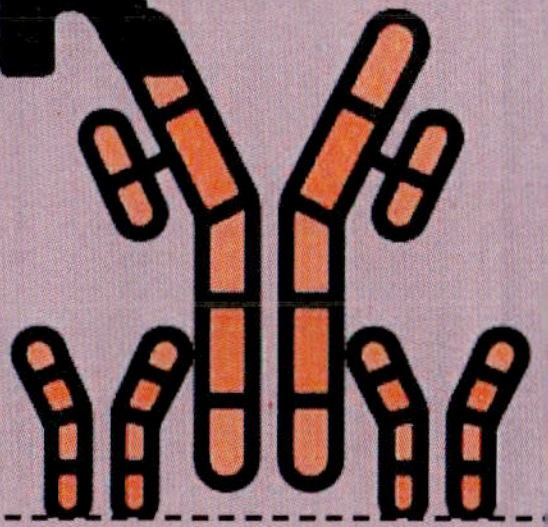

UNIQUE PROTEINS ON PATHOGENS

EPITHELIAL

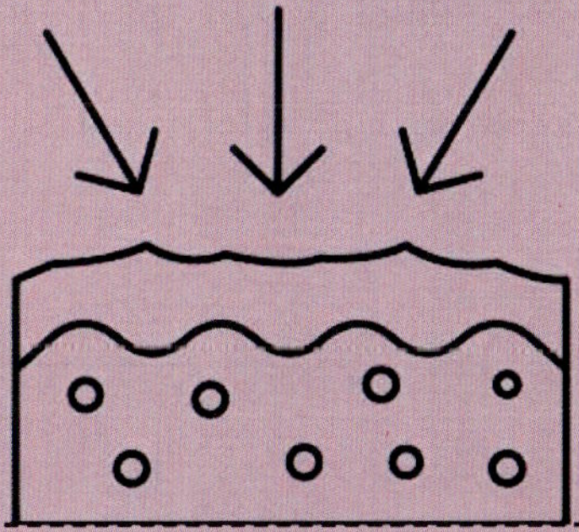

OUTER LAYER

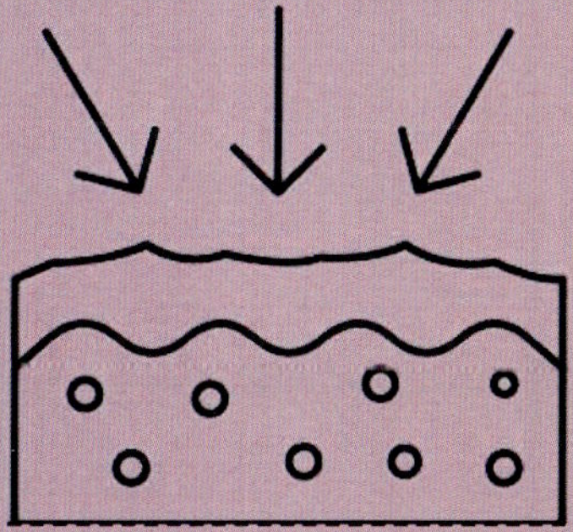

INTEGUMENTARY

SKIN, HAIR, & NAILS

BIOCHEMISTRY

BIOLOGICAL CATALYSTS

ENZYMES

SYNTHESIZE

PUT TOGETHER

ASSEMBLE

SYNTHETIC

ARTIFICIAL

MAN-MADE

FAKE

IMITATION

BASIC

ALKALINE

COHESIVE

LIKE THINGS
STICK
TOGETHER

ADHESIVE

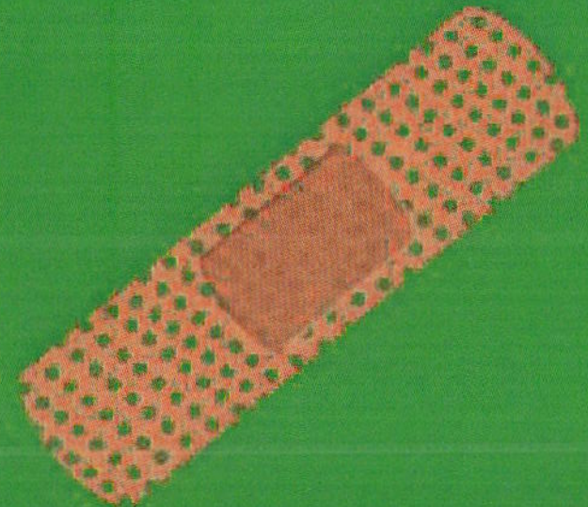

UNLIKE
THINGS
STICK
TOGETHER

CONCENTRATED

A LARGE
QUANTITY OF A
SUBSTANCE IN
AN AREA

DILUTE

A SMALL QUANTITY OF A SUBSTANCE IN AN AREA

OPTIMUM

MOST/ HIGHEST

EXOTHERMIC

ENERGY RELEASING

ENDOTHERMIC

ENERGY ABSORBING

ALBUMIN

EGG WHITE

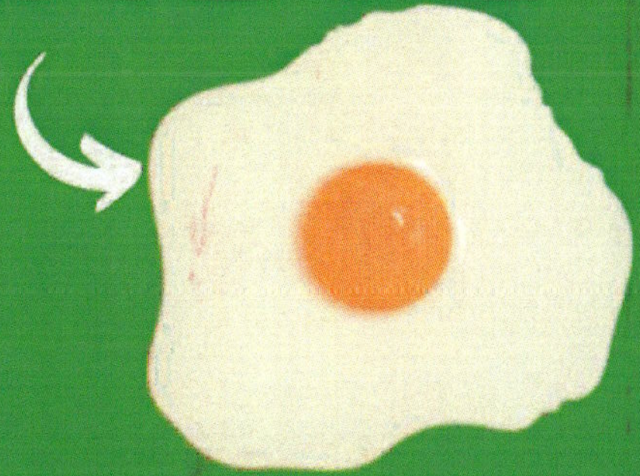

DENATURE

CHANGE SHAPE

ALTER

PROTEIN

POLYPEPTIDE

AMINO
ACID
CHAIN

POLYPEPTIDE
CHAIN

CELLULAR
ENERGY

ATP

ADENOSINE
TRI-PHOSPHATE

ENERGY

CHLOROPLAST

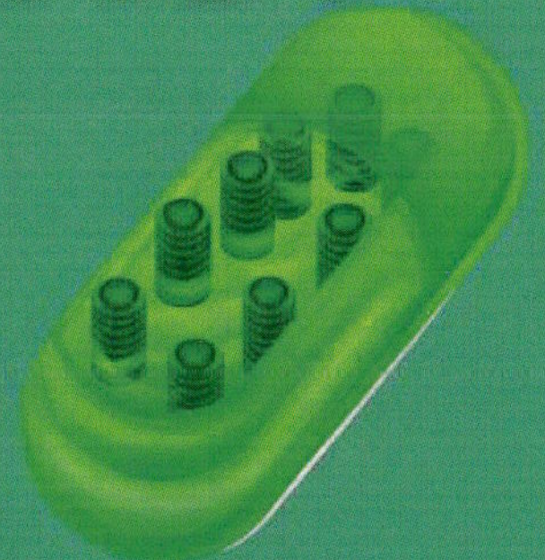

ORGANELLE
THAT CONTAINS
CHLOROPHYLL
& DOES
PHOTOSYNTHESIS

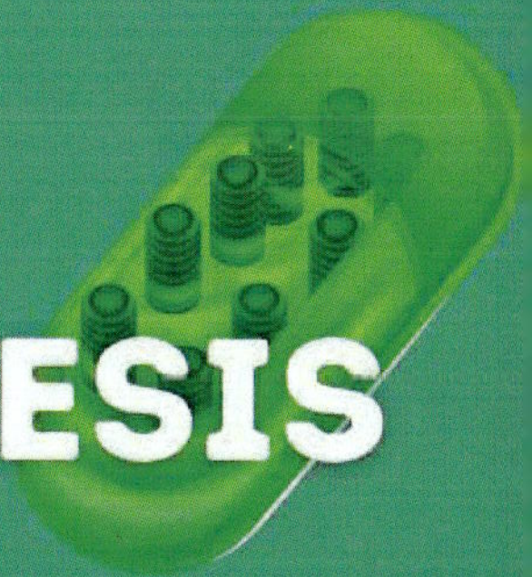

RESPIRATION

CELLULAR
RESPIRATION

AEROBIC RESPIRATION

RESPIRATION USING NUTRIENTS & OXYGEN TO MAKE ENERGY

ANAEROBIC RESPIRATION

FERMENTATION

USING NUTRIENTS TO MAKE ENERGY IN THE ABSENCE OF OXYGEN

PHOTOSYNTHESIS

PROCESS USED BY AUTOTROPHS TO MAKE NUTRIENTS USING LIGHT

CHEMOSYNTHESIS

PROCESS USED BY CERTAIN ORGANISMS TO MAKE NUTRIENTS WITH CHEMICALS IN THE ABSENCE OF LIGHT

DICHOTOMOUS

OPPOSING

CONTRADICTORY

HIERARCHY

RANKINGS

ORGANIZED IN A PYRAMID OF LEVELS

HETEROTROPH

MUST TAKE IN NUTRIENTS

CONSUMER

AUTOTROPH

MAKES ITS OWN NUTRIENTS

PRODUCER

PROKARYOTE

CELLS DO NOT HAVE NUCLEUS

EUKARYOTE

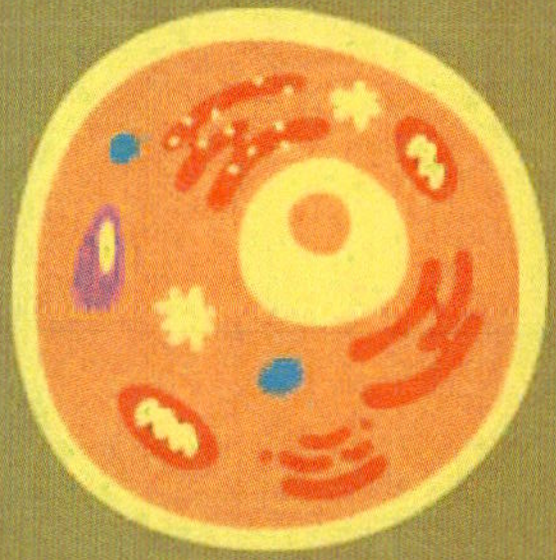

CELLS HAVE A NUCLEUS

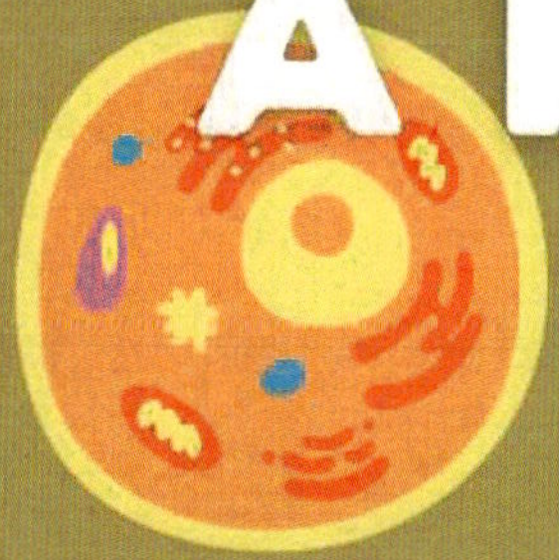

ARCHAEBACTERIA

ANCIENT MICROORGANISMS

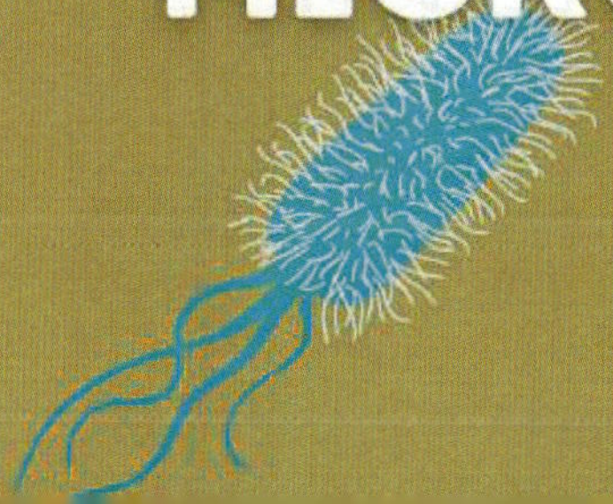

CYANOBACTERIA

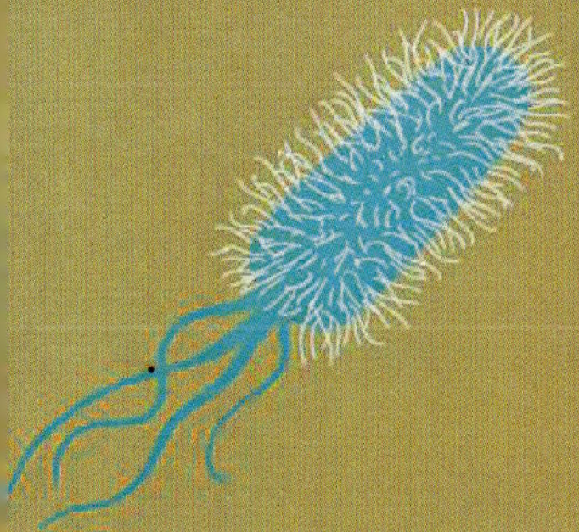

BLUE-GREEN BACTERIA

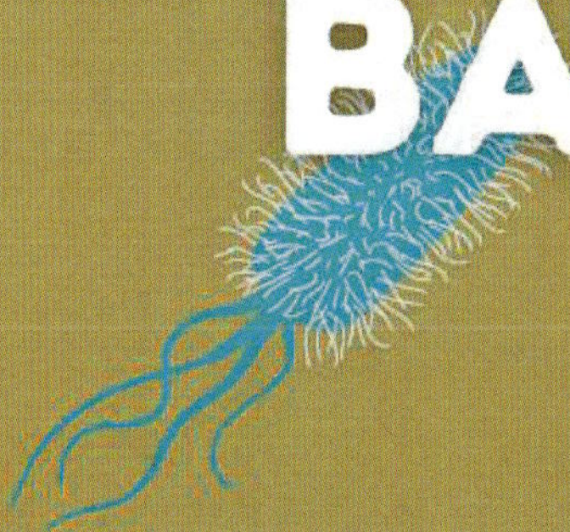

EUBACTERIA

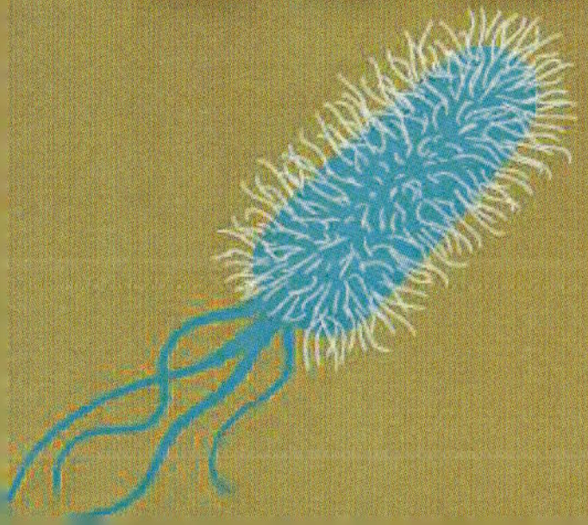

TRUE BACTERIA

MONERA

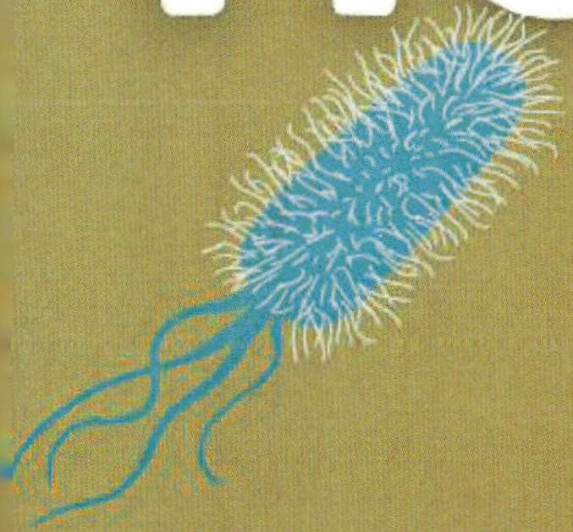

J-SHAPED

EXPONENTIAL

DIRECT

CONIFEROUS FOREST

BOREAL FOREST

TAIGA BIOME

AQUATIC
BIOME

MARINE/
OCEAN
BIOME

FRESH
WATER
BIOME

TERRESTRIAL

LAND

ENERGY PYRAMID

ECOLOGICAL PYRAMID

BIODIVERSITY